Nadjet BENADLA
Mohammed DEBBAL

Design and implementation of a mobile telemedicine system

Nadjet BENADLA
Mohammed DEBBAL

Design and implementation of a mobile telemedicine system

Wireless transmission of ECG signals

Imprint

Any brand names and product names mentioned in this book are subject to trademark, brand or patent protection and are trademarks or registered trademarks of their respective holders. The use of brand names, product names, common names, trade names, product descriptions etc. even without a particular marking in this work is in no way to be construed to mean that such names may be regarded as unrestricted in respect of trademark and brand protection legislation and could thus be used by anyone.

Cover image: www.ingimage.com

This book is a translation from the original published under ISBN 978-620-6-70585-7.

Publisher:
Sciencia Scripts
is a trademark of
Dodo Books Indian Ocean Ltd. and OmniScriptum S.R.L publishing group

120 High Road, East Finchley, London, N2 9ED, United Kingdom
Str. Armeneasca 28/1, office 1, Chisinau MD-2012, Republic of Moldova, Europe
Printed at: see last page
ISBN: 978-620-7-27772-8

Table of contents

GENERAL INTRODUCTION

Mobile telemedicine systems are becoming increasingly important, particularly in the care of patients who are isolated or traveling, far from a referral hospital. These systems need to be cost-effective, small and have low power consumption. They need to be manageable and easy for patients to use.

The incorporation of technologies such as Bluetooth and GPRS enables wireless transmission to health centers.

Figure 1 shows the schematic diagram of wireless transmission from the patient to the doctor. The patient, equipped with the acquisition card, transmits his cardiac status using his cell phone. The mobile phone, connected to the GPRS network, can then communicate with the hospital. The doctor, for his part, receives the information on his computer. He studies the severity of the treatment and, depending on the patient's condition, can either consult directly by telephone or intervene in an emergency.

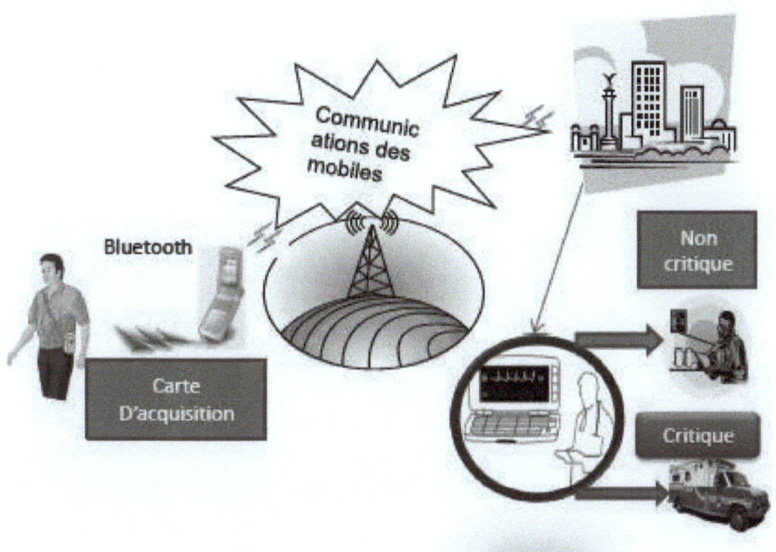

Figure.1 Our project scenario [1].

For our project, we are interested in the realization of the ECG signal acquisition board. This work has been developed in our faculty's electronics laboratory.

Our dissertation consists of four chapters, preceded by an introduction. It concludes with an outlook.

Chapter 1 introduces the electrical functioning of the heart, electrocardiography and the electrocardiogram. While the various JAVA platforms, particularly J2ME and its configurations and profiles, as well as its basic CLDC and MIDP libraries, are described in Chapter 2. Chapter 3 contains samples and explanations of the JABWT programming used in the development of client-server applications. It covers the basic operations of client and server initialization and discovery.

Finally, Chapter 4 is devoted to the practical realization of the system, or the essential components used. These are the TL084 for amplification, the PIC 16F877A for analog/digital conversion, and the Bluetooth F2M03GX module for transmission to the cell phone.

CHAPTER I
ELECTROCARDIOGRAPHY (ECG)

I.1 Introduction

The heart plays a vital role in the human body [3]. It ensures the movement of blood in the small circulation (VD-OG) and the large circulation (VG-OD). This performance depends on an original anatomical organization in which the heart's chambers communicate via different orifices equipped with anti-backflow valves, an automatic excitation system, a particular contractile mechanics, and finally a correct supply of energy and oxygen [4]. To this end, recent technological advances have developed heart signal detection devices known as electrocardiography. In this chapter, we present a review of the literature on the electrical functioning of the heart, electrocardiography, the electrocardiogram, and the study and analysis of ECG signals.

I.2 Electrical functioning of the heart

As with all muscles in the body, myocardial contraction is caused by the propagation of an electrical impulse along the cardiac muscle fibers, induced by depolarization of the muscle cells. In the heart, depolarization normally originates at the top of the right atrium (the sinus), and then propagates through the atria, inducing atrial systole (Figure I.1), followed by diastole (muscle relaxation). The electrical impulse then arrives at the atrioventricular (AV) node, the only possible crossing point for the electrical current between the atria and ventricles. Here, the electrical impulse pauses briefly, allowing blood to enter the ventricles. It then passes through the His bundle, which is made up of two main branches, each going into a ventricle. The fibers making up this bundle, supplemented by Purkinje fibers, thanks to their rapid conduction, propagate the electrical impulse to several points in the ventricles, and thus enable almost instantaneous depolarization of the entire ventricular muscle, despite its large size, ensuring optimum efficiency in blood propulsion; this contraction constitutes the "His bundle".
the ventricular systole phase. This is followed by ventricular diastole (muscle relaxation), when muscle fibers re-polarize and return to their initial state [5].

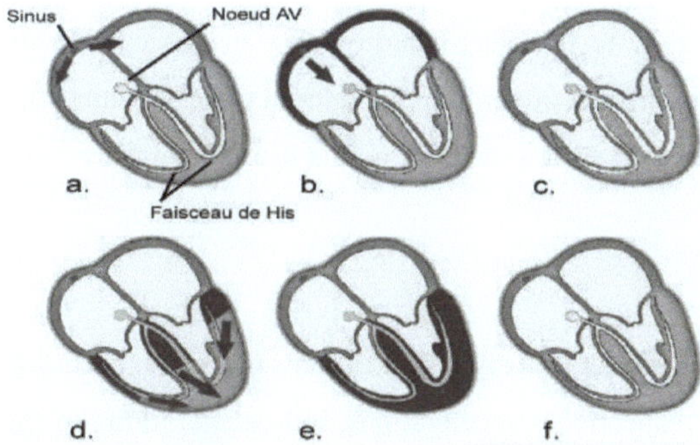

Figure I.1: Electrical operation of the core [1].

I.3 Electrocardiography

Electrocardiography explores the heart's electrical activity by recording electrocardiograms (ECGs), graphs showing differences in electrical potential at different points in the body as a function of time.

These recordings are made possible by the very nature of cardiac activity: cardiac muscle fiber activation can be divided into two phases: depolarization and repolarization. The depolarization phase is very abrupt, in a way the main phase; the repolarization phase is designed to restore charges to their initial values. Induced potential differences are only of the order of a few millivolts, but they are large enough to be detected in the human body, which is a fairly homogeneous conducting medium. As a result, several electrodes placed at different points on the body do not perceive the same electric current.

There are several ways of performing an electrocardiogram. The first was defined by Einthoven, inventor of the electrocardiograph, who won the Nobel Prize for his discovery in 1924.

Three electrodes are placed, one on each wrist and one on the left foot, when the subject is lying down, with the heart at the approximate geometric center of the triangle. This arrangement enables easy vector calculations, as each lead (left wrist - right wrist, left wrist - left foot and right wrist - left foot) constitutes an electrical circuit (electrode 1, wire from electrode to electrocardiograph, wire from electrode 2 to electrocardiograph, human body).

It is also possible to use so-called unipolar leads, where the electrodes are not connected to each other, but only to the electrocardiograph. In this case, more electrodes are needed to obtain the electrocardiogram, and they are implanted on both wrists and the left foot, as well as at a series of points forming an arc around the heart.

I.4 Electrocardiogram

Electrocardiograms can easily detect rhythm disorders such as tachycardias and bradycardias, which are respectively accelerations and decelerations of the heartbeat, as well as arrhythmias. The electrocardiogram also enables us to study atrial and ventricular waves in much greater detail. Atrial waves, known as P waves, enable us to check whether the atrial contraction frequency, which must precede that of the ventricles by around 16 hundredths of a second, has been respected. Ventricular waves, known as T waves, are used to check that the electrical activity of the ventricles, the heart's masterpieces since they expel blood into the vessels, has been respected. It should not be forgotten, however, that electrocardiography can also be used to detect pulmonary or metabolic problems [6].

I.5 Core electrical trace

The heartbeat can therefore be tracked by surface recording of the electrical signal that accompanies it. In fact, each phase of the beat has a specific electrical trace. A trained eye can therefore, in most cases, quickly differentiate the trace of an atrial contraction from the trace of a ventricular contraction. Let's apply the ECG principle. To the electrical activity of a normal heartbeat. The initial impulse comes from the sinus: it's not visible on the ECG. The electrical wave that then propagates through the atria, causing them to contract, leaves the trace of a small positive deflection on the ECG: the P wave (Figure I.2a). The impulse then arrives at the atrioventricular (AV) node, where the short pause is reflected on the ECG as a small flat segment; it then travels along the fast conducting pathways (the His bundle) to cause contraction of the ventricles, followed by their repolarization.

This propagation of the impulse, and the brief, powerful contraction of the entire ventricular muscle, creates a succession of 3 waves (Q, R and S) on the ECG, known as the QRS complex (Figure I.2b). The Q wave is the first: it's a downward-facing wave, not always visible on the trace; the second is the R wave: it's large in amplitude and upward-facing; the last is downward-facing: it's the S wave. After each QRS complex, a wave called the T wave appears on the ECG. Between this wave and the previous one, there's a short pause called the ST segment, the study of which is very important for identifying certain pathologies. The T wave reflects the repolarization phase of the cells making up the ventricles; this is a purely electrical phenomenon, during which the heart is mechanically inactive (Figure I.2c) [6].

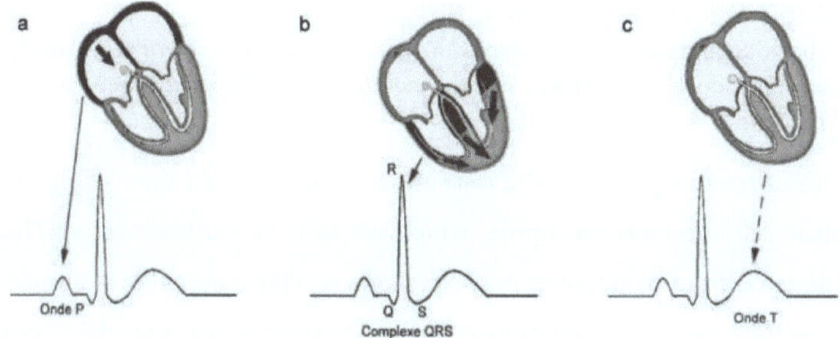

Figure I.2: Core electrical trace [1].

I.6 Study and analysis of ECG signals
I.6.1 Physiological origins

Electrical potentials originate in the fibers of the heart muscle. The generation of excitation in the various parts of the heart can be studied not only by measuring the electrical potentials of the cells or the electrical potentials on the surface of the heart, but also by recording cardiac activity at skin level. Indeed, as potential differences develop between excited areas of the heart, differential electrical forces propagate throughout the body. Tracings reflecting oscillations in these potentials can therefore be recorded by applying electrodes to certain points on the body. In a simplified model, the heart, which is the source of the signals, is a generator represented by an electrical dipole located in the thorax.

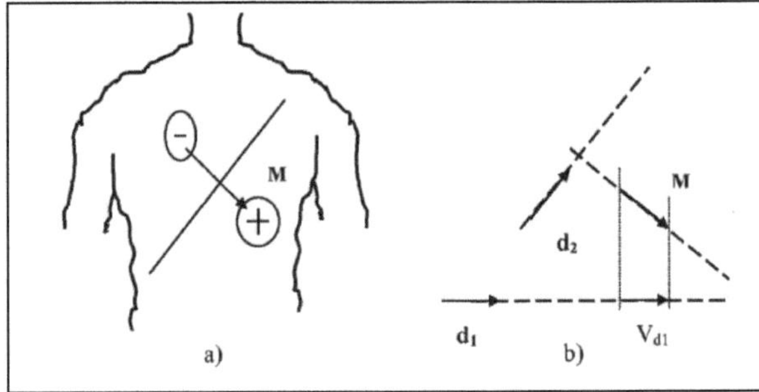

Figure I.3: The cardiac vector **[2].**
   a) The dipole results from positive and negative charges separated from each other; the moment vector is represented by M.
   b) The voltage measured in a derivation represented by the vector $\mathbf{d_1}$ is simply the scalar product $V_{d1} = \mathbf{M.d_1}$; it is the modulus of the projection of $\mathbf{M}$ in the direction $\mathbf{d_1}$.

In electrocardiography, we're not interested in the field lines of this dipole, but instead use the moment dipole, which we call the cardiac vector. This is a vector directed from negative to positive charges, whose modulus is proportional to the quantity of charges multiplied by the distance separating the 2 types of charge. During a cardiac cycle, the amplitude and direction of this vector vary. The potential difference measured by 2 electrodes represents the modulus of the projection of the cardiac vector on the straight line connecting these two electrodes.

By convention, an electrical impulse propagating towards the electrode is represented on the electrocardiogram recording by a deflection towards the top of the trace. If, on the other hand, the electrical activity flees the electrode, a downward deflection is observed. Figure I.4 shows the paths taken by the impulses in the heart **[3].**

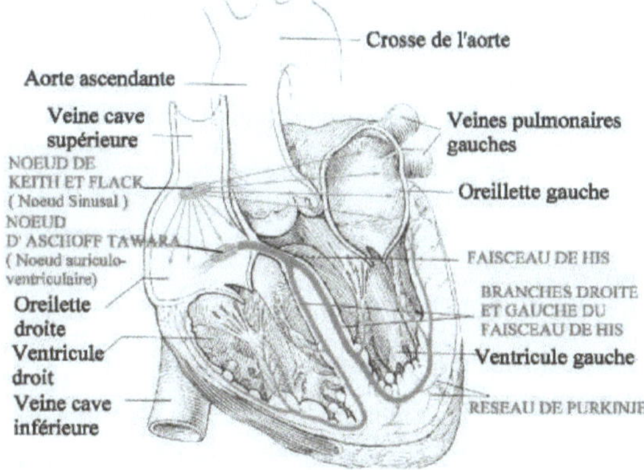

Figure I.4: Conduction circuit for electrical excitations in the heart [3].

I.6.2 Deviations

There are different bypass systems for positioning and connecting electrodes to the device.

The standard ECG is made up of 12 separate leads: six thoracic leads and six limb leads. These leads study the projection of cardiac activity to the body's periphery in two planes: frontal and horizontal.

a. EINTHOVEN bipolar limb diversion

These leads use 3 electrodes placed on the subject. The electrodes are placed on the right and left arms and on the left leg to form a triangle (Einthoven triangle).

These leads are called bipolar because they measure a potential difference between two electrodes.

Each side of the triangle formed by the three electrodes represents a lead (DI, DII, DIII), using a different pair of electrodes for each lead.

DI: right arm electrode connected to the negative pole of the galvanometer, left arm electrode connected to the positive pole. (DI records the potential of the left arm (VL) minus the potential of the right arm (VR).

DII: right arm electrode connected to negative pole, left leg electrode connected to positive pole. (DII registers VF - VR).

11

DIII: left arm electrode connected to negative pole and left leg electrode to positive pole. (DIII registers VF -VL).

By translating these three derivations towards the center of the triangle, we obtain the intersection of three reference lines.

b. Unipolar limb derivation

According to Kirchhoff's 2nd law, the sum of the absolute potentials recorded in the right arm, left arm and left leg is equal to zero. Thus, VR +VL +VF = 0. From this, it would be possible to obtain an indifferent electrode that would remain at zero potential by connecting the three ends through equal resistors (5000 ohms) to the center. By connecting the center to the negative pole of the electrocardiograph and an exploratory electrode to the positive pole, we obtain a system in which the exploratory electrode records potential variations above and below zero potential. In this way, from the right arm, left arm and left leg, the following leads are recorded

are designated VR, VL and VF. The recording obtained by these leads is weak, and Goldberger suggested that the potential at one end could be increased by disconnecting its center electrode and then recording the potential difference between that end and the other two electrodes.

This unipolar lead completes the first. The ECG records potential variations at the level of the exploratory electrode (placed at a given (fixed) point on the body surface): VR = right arm, VL = left arm, and VF = left leg. The other "indifferent" electrode, connected to zero potential, corresponds to the heart's electrical center.

To obtain a trace with the same amplitude as leads DI, DII and DIII, the ECG voltage had to be amplified (increased). The result is lead A (augmented), V (voltage), R (right arm), and two additional leads:

* AVR: We record the potential difference between the right arm (BD) and the middle BG-JG. This lead uses the right arm as a positive and all other limb electrodes as a negative (common) ground.

* AVL: uses left arm as positive

* AVF: tension of the leg in relation to the two arms used as reference. The positive electrode is located on the left leg.

These leads reflect electrical activity in the region of the heart opposite the electrode. VF reflects electrical activity on the underside of the heart. VL reflects the electrical activity of the upper part of the left side, and VR that of the ventricular chambers.

The three successive derivations are projected onto the three bisecting lines, which intersect at the center. These AVR, AVL and AVF derivations intersect at different angles, producing the intersection of three other reference lines.

The six leads DI, DII, DIII, AVR, AVL and AVF join to form six reference lines that intersect precisely and lie in a frontal plane on the subject's thorax.

Each limb lead records from a different angle, so each represents a different view of the same cardiac activity.

These leads allow vector calculations, since each 'sees' the same cardiac activity but from different angles s

c. Precordial unipolar shunt

The active (positive) electrode is placed at different levels of the thorax.

Thoracic leads are designated V1 to V6. They run progressively from the right to the left of the subject. These thoracic leads project through the nd AV towards the patient's back, which is the negative pole of each thoracic lead.

Leads V1 and V2 are placed opposite the right chambers of the heart and therefore explore the septum and right ventricle; they reflect right ventricular activity.

Leads V3 and V4 are located opposite the interventricular septum. They explore the tip of the heart and the anterior surface of the left ventricle.

V5 and V6 face the left cavities and explore the lateral wall of the left ventricle.

As they are not equidistant from the heart, these leads do not allow vector analysis.

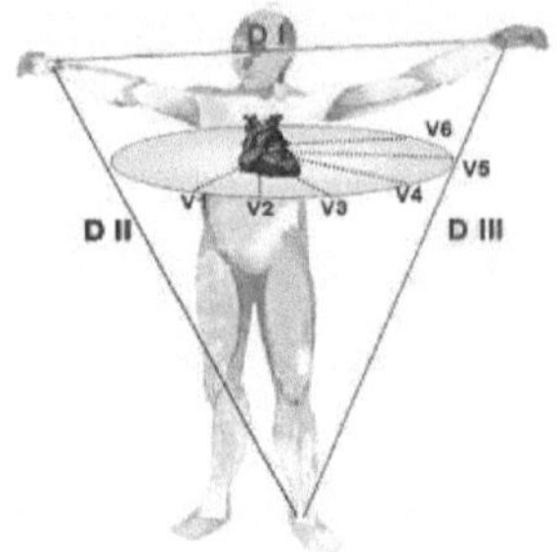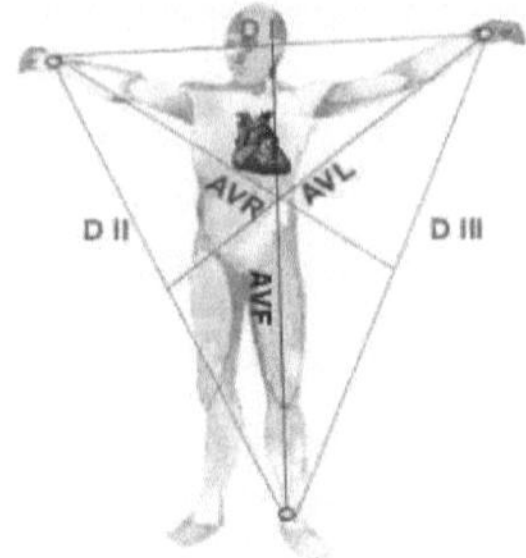

Figure I.5: Standard leads DI, DII, DIII, unipolar aVR, aVL, aVF and precordial V1, V2, V3, V4, V5, V6 [4].

d. morphology of the electrocardiographic trace

The P wave is a rounded accident. It has a low amplitude of 0.1 to 0.2 mV, corresponding to depolarization of the atria. This wave occurs just before OR contraction. (Ionic phenomena always precede electrical phenomena). It is generally positive.

In dogs and horses, splitting is quite common. In humans, it is pathological.

Repolarization of the atria is masked and goes unnoticed.

The QRS complex witnesses ventricular depolarization. Observed before ventricular systole.

None of the QRS events has any particular explanation; the entire ventricular mass is not activated at the same time, but progressively.

The T wave reflects the repolarization of the ventricle. It is generally on the same side as the QRS event, as depolarization takes place from the inside to the outside of the heart (endocardium - epicardium).

In the human heart, repolarization is slow and takes place in the opposite direction to depolarization (epicardium - endocardium); as the currents recorded are in the opposite direction to those of repolarization (the current going from cells still depolarized with a negative exterior to cells already repolarized with a positive exterior), it follows that the depolarization wave (QRS) and the repolarization wave (T) are in the same direction. During ECG recording with an exploratory electrode placed on the epicardial surface: at rest and when the entire thickness of the wall is depolarized, no current is produced, and consequently, no electrical

14

activity appears. Current emission occurs when opposite charges are in contact during depolarization and repolarization.

In birds, the distribution of excitatory tissue is mainly distributed over the epicardium, so the T wave is rotated in the opposite direction to R (P is masked, high frequency above 300. QRS = single peak = R).

The PQ interval (PR) is the time between the depolarization of the OR and the depolarization of the Vent. This PQ time encodes what is known as = atrioventricular conduction. It is isoelectric.

The QT interval is the time during which the ventricle is subjected to variations in electrical phenomena known as electrical systole.

By analyzing the details of potential variations, the trained person obtains valuable information relating to: the anatomical orientation of the heart in the thorax; the size of the cardiac chambers; rhythm and conduction disorders; the extent, location and progress of myocardial ischemic conditions; the effects of altered electrolyte concentrations in the ECF, etc.

On the other hand, it should be remembered that the ECG does not provide direct information about the mechanical functioning (performance) of the heart as a pump. For example, valvular insufficiency will not be detected unless accompanied by abnormal conduction **[4]**.

I.7 Remote medical monitoring

Medical telemonitoring is defined as remote monitoring based on a global information system comprising people equipped with physiological sensors linked in networks that may be wireless or wired for real-time collection of patient data, automatic devices for adapting the person's living environment to his or her personal abilities (motor and cognitive); a local unit, at the level of each person, and also for processing signals received from the sensors, managing a knowledge base relating to the patient. At the origin of messages and alarms; a tele-vigilance center for processing messages and alarms received from individuals, as well as a set of actors (medical staff, patient and family members) who can access the system's data at any time, after authentication and according to their privileges.

Information and communication technologies **are**   also changing the data, and it's thanks to these technologies that the patient can be directly connected to the medical team.

The advantages and disadvantages of remote medical monitoring are as follows

Advantages include

- The person can live at home for as long and as independently as possible, in an environment of comfort and safety.
- Prevent health deterioration as early as possible.
- Save on emergency medical transport and advanced hospitalization.
- Improving patient communication with healthcare professionals, caregivers and family members.
- Reduced hospitalization costs.
- Easy access to data, i.e. reference to blood glucose, blood pressure, heart rate and oxygen levels, as well as medical advice.
- By connecting to a monitor to view live data from all patients, this enables nurses to perform flexible monitoring functionality at the point of care, resulting in more time spent with patients, which translates into better quality of care.

The disadvantages of telemonitoring are that, until now, communication between patient and medical staff has been by telephone, and these are the only relationships that can ensure patient safety at a distance.

So, if it's not feasible to undertake home treatment without a telephone. This is likely to result in insufficient telephone contact, due to the subjectivity of the dialogue. The number of parameters actually transmitted, as well as the random frequency of calls, may be limited, and the availability of care teams often variable. These shortcomings, in both the quality and quantity of telephone relations, can lead to problems to which the care teams are accustomed. These problems often result in the unexpected transfer of the patient to the follow-up center for diagnosis, and sometimes hospitalization when the call is too late.

Given that the transmission of raw medical data imposes a heavy workload on doctors, a workload that can inevitably limit the number of patients treated in this way.

Apart from all the above, there is also the lack of human contact, which is a kind of therapy for the patient, who needs a lot of affection [7].

I.8 Conclusion

Analysis of such recordings requires the use of automatic signal reading tools, as the quantity of information recorded in 24 hours is very large: it corresponds to around 100,000 heartbeats over 3 recording channels [8].

To facilitate patient monitoring, we have proposed a portable device for ECG signal detection using wireless transmission technologies and J2ME programming. Chapter II describes J2ME.

CHAPTER II
JAVA2 MOBILE EDITION (J2ME)

II.1 Introduction

The first cell phones were designed to call only, so their displays were as simple as possible. Over time, technological evolution has made phones more and more powerful, with more memory and better displays.

Today, almost all phones are equipped with a camera, mp3 player or radio. However, phones are evolving, but to standards that differ from manufacturer to manufacturer and from model to model. Application development generally involves the use of a proprietary API, often written in *C* or *C++*. Portability of an application therefore implies adapting the code for almost every phone model.

Sun offers a lighter version of J2SE (Java 2 platform Standard Edition), adapted to low-power devices, called J2ME **[10]**.

In this chapter, we present the essential functionalities and various components of this mobile application development environment.

II.2 Description of J2ME

J2ME (figure II. 1) is a subset of the J2SE (Standard Edition) platform. It aims to retain the qualities of Java technology by adopting the architecture of the other editions, i.e. a virtual machine, native libraries and an API, while retaining multi-product consistency, secure network operation and bottom-up extensibility.

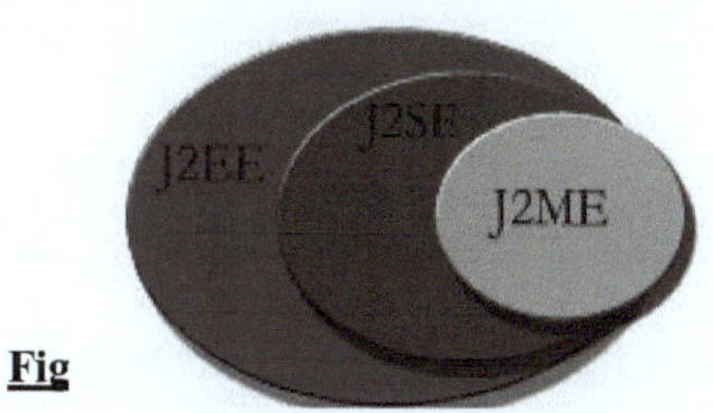

Fig

II.3 The diversity of peripherals

Setting up a common technical platform for all peripherals is a very delicate operation. Each device has its own specific features, requiring applications to adapt to different display or pointing characteristics.

The following list shows the variety of terminals targeted by Sun :

- Cell phone, Smartphone (Nokia, Ericsson, Alcatel, Siemens, ...) ;
- PDA (Palm Pilot, PocketPC, ...) ;
- Digital imaging devices (digital camcorders, digital cameras, etc.) ;
- Automatic industrial devices (Robot in a machining line, On-board display in automobiles...).

II.4 J2ME architecture

As mobile terminals don't have the same resource capacities as conventional desktop computers (memory, disk and computing power), they require a leaner environment, adapted to different execution constraints (figure II.2).

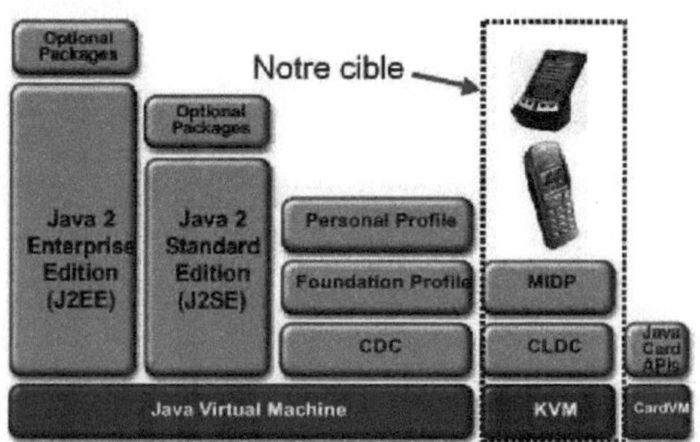

Figure II.2: J2ME platform architecture **[11]**.

As terminals don't have the same resource capacities (memory, disk and computing power) as traditional desktop computers, the solution involves providing a lighter environment to adapt to different execution constraints. However, how can we integrate the diversity of our offer into a technical base whose target is not defined a priori? The solution proposed by J2ME consists in grouping certain product families into categories, while offering the possibility of implementing specific routines for a given terminal. The strengths of this solution lie in the richness of its user interface, and in its ability to operate in either connected or disconnected mode.

The J2ME architecture is therefore divided into several layers with different technical constraints:

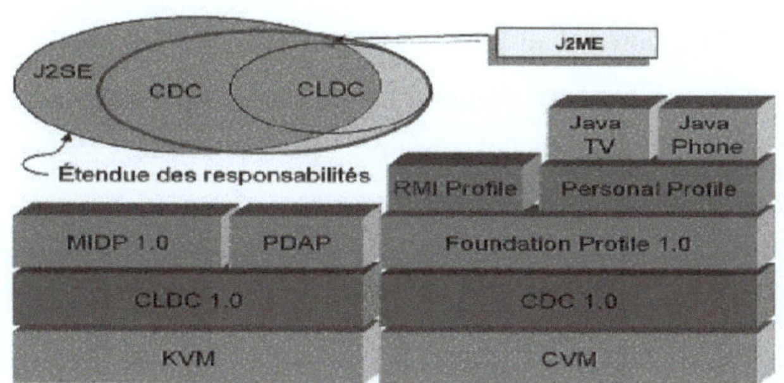

Figure II.3: Different J2ME layers **[12]**.

The aim of this layered architecture (figure II.3) is to factorize a set of APIs for given product families, enabling an application to run on several terminals without code modification.

The basic building blocks of J2ME are configuration, profile and optional packages **[12]**.

A configuration is a virtual machine and a minimal set of base classes and APIs. It specifies a generalized execution environment for embedded terminals and acts as the Java platform on the terminal;

A profile is a Java API specification defined by the industry and used by manufacturers and developers for different types of terminal;

An optional *package is*, as its name suggests, a package that may not be implemented on a particular terminal.

The virtual machine sits between the application and the platform used, converting the application's code bites into the machine mode appropriate to the hardware and operating system in use. Depending on the target, the virtual machine can be slimmed down to consume more or less resources. J2ME currently offers 2 types of virtual machine, KVM (KiloByte Virtual Machine) and CVM (Convergence Virtual Machine).

The Java classes running in the KVM have been designed to operate in a limited environment in terms of memory (128 K), power and network access. This configuration is therefore designed to save resources, with a 40 to 80 Kbyte KVM running 30 to 80% slower than a normal JVM.

As for CVM, it has been designed for terminals requiring the JVM's full set of functionalities, but with smaller capacities. Terminals using CVM are generally compact, connected and consumer-oriented.

II.4.1 Configurations

They define a minimum platform in terms of services for one or more given profiles. Two configurations are proposed today: CDC (Connected Device Configuration) and CLDC (Connected Limited Device Configuration) [13].

- The CDC specifies an environment for high-capacity connected terminals such as set-top boxes, display telephones and digital TV. The characteristics of the hardware environment proposed by the CDC configuration are :

- minimum 512KB ROM and 256KB RAM, 32-bit processor;

- a mandatory network connection (wireless or not) ;

- support for the complete Java Virtual Machine (CVM) specification.

This configuration is part of an almost complete Java architecture.

- CLDC targets resource-constrained devices such as cell phones, PDAs and lightweight wireless devices. As these devices are limited in terms of resources, the conventional environment is unable to meet the memory occupancy constraints associated with these devices. J2ME has therefore defined a set of CLDC-specific APIs designed to take advantage of the specific features of each terminal in a given family (profile). The characteristics of the hardware environment proposed by the CLDC configuration are :

- A minimum of 160KB to 512KB RAM, 16 or 32-bit processor, 16 MHz speed or higher;

- Limited power supply, battery support;

- A non-permanent, wireless network connection;

- Limited or no graphical user interface.

However, CLDC does not integrate the management of graphical interfaces, persistence or the particularities of each terminal. These aspects are not CLDC's responsibility.

II.4.2 Profiles

They enable a certain category of terminals to use common features such as display management, input/output events (pointing, keyboard, etc.), or persistence mechanisms (integrated lightweight database). These profiles are subject to specifications based on the JCP (Java Community Process) principle.

Profiles define the complete set of API classes that will be made available to a J2ME application and designed specifically for a given configuration (figureII.4).

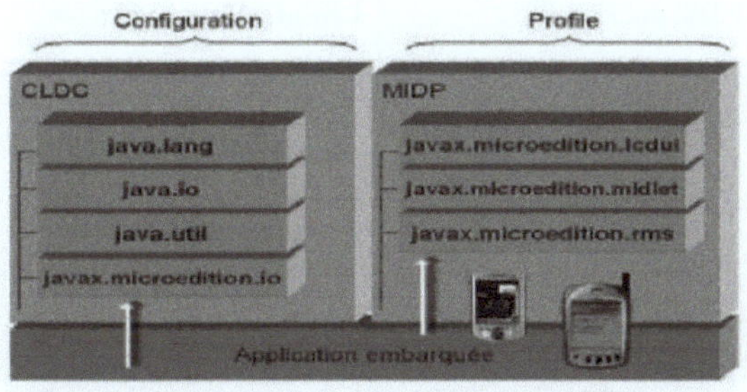

Figure II. 4: Configuration, J2ME Profile **[14].**

Sun offers two J2ME reference profiles: the Foundation profile and the Mobile Information Device Profile (MIDP).

The Foundation profile is intended for CDC configuration and therefore offers all the richness of a virtual machine virtually identical to the standard virtual machine. This means that developers using the Foundation profile have access to a full implementation of J2SE functionality **[14]**.

MIDP (Mobile information device profile)

MIDP is the basis for implementing the classes linked to a given profile. These include methods for display, user input and persistence (database) management. Today, there are two major implementations of MIDP profiles. One is more specific, designed for Palm Pilot-type assistants (PalmOs), and the other is totally generic, proposed by Sun as a reference implementation (RI).

These profiles are freely downloadable from Sun's website, and include several emulators for software-based application testing.

II.5 The J2ME API

The J2ME API is divided into two parts, one specific to MIDP and the other to CLDC, whose lists of packages are illustrated in Table II.1. We will now look at these two parts in detail.

List of CLDC packages	List of MIDP packages
java.io	javax.microedition.lcdui
java.lang	javax.microedition.midlet
java.util	javax.microedition.rms
javax.microedition.io	

Table II.1: J2ME APIs

II.5.1 API CLDC 1.0

As CLDC is a lightweight java environment, the libraries it provides are reduced in size, but nevertheless

a. java.io

sufficient for the development of applications on targeted terminals.

The *java.io* package provides input/output flow management, containing the classes and methods needed to retrieve information from remote systems.

b. java.lang

The *java.lang* package is a subset of the standard J2SE *java.lang* package classes, such as the Runnable interface or the Boolean, Integer, Math, String, Thread, Runtime and other classes (15 in all).

c. java.util

The *java.util* package contains a small subset of the corresponding J2SE package, including the following classes: Calendar, Date, TimeZone, Enumeration, Vector, Stack, Hashtable and Random.

d. javax.microedition.io

This library contains classes for connecting via TCP/IP or UDP. The main object in the *javax.microedition.io* package is the Connector class. This class provides a uniform and convenient way of performing input/output regardless of protocol type.

II.5.2 API MIDP

As MIDP is the profile associated with CLDC configuration, it supports the highest-level functionality [15].

a. MIDP version 1

MIDP version 1.0 a is the result of work carried out by the Java Community Process Expert Group, JSR 37 [16].

MIDP adds a few packets to the CLDC as shown in Table II.2.

Package	Supplies

javax.microediton.lcdui	Provides classes for the user interface
javax.microedition.midlet	Defines MIDP applications and the interactions between the application and the environment
javax.microedition.rms	Provides persistent storage (system register management)

Table II.2: MIDP 1.0 packages

b. *MIDP version 2*

MIDP version 2.0 is the result of the Java Community Process Expert Group, JSR 118. The MIDP 2.0 specification defines an augmented architecture and associated APIs for developing applications for mobile information devices.

The specifications are based on those of MIDP 1.0, providing compatibility so that MIDlets written for MIDP 1.0 can run in MIDP 2.0 environments.

The only difference is the increase in memory in the MIDP 2.0 specifications. Table II.3 shows the packages provided by MIDP 2.0.

Package	Supplies
javax.microediton.lcdui	This library provides all the classes needed to manage the terminal user interface
javax.microedition.midlet	Defines MIDP applications and interactions between the application and the environment in which it operates (midlet lifecycle management).
javax.microedition.rms	Contains all the classes needed to manage a lightweight database on a terminal.
javax.microedition.lcdui.game	Provides useful functionality for game development
javax.microedition.media	Provides audio module
javax.microedition.pki	Provides functionality for certificate handling

Table II.3: MIDP Packages

II.6 Midlets

Applications created with MIDP are midlets: these are classes that inherit from the abstract class *javax.microedition.midlet.Midlet*. This class enables dialogue between the system and the application.

It has three methods for managing the application's life cycle according to three possible states: active, suspended or destroyed (figure II.5). The first method, **startApp()**, is called each time the application is started or restarted. The second, **pauseApp()**, is called when the application is paused. Finally, the third method, **destroyApp()**, is called when the application is destroyed. Note that the life cycle of a midlet is similar to that of an applet [17].

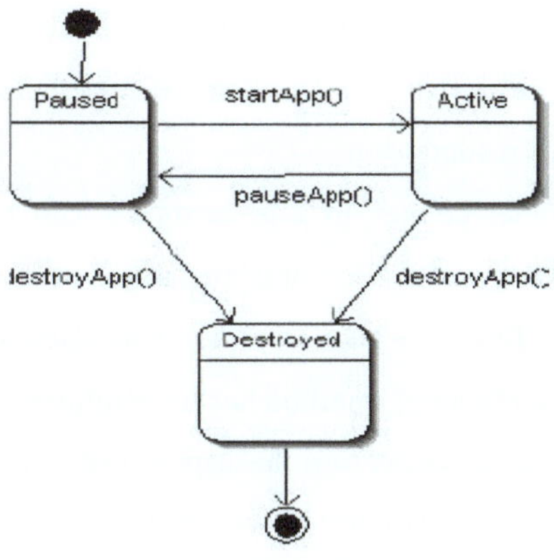

Figure II.5: Midlet life cycle [13].

II.7 The user interface

MIDP is designed to run on many different types of terminal: telephones, Palm Pilots, etc. While respecting these diversity constraints, MIDP applications must always integrate the same functionalities, whatever the terminal. The solution was to break down the user interface into two layers: the high-level API and the low-level API. The former favors portability, the latter the exploitation of all terminal functionalities. The designer must therefore make a compromise between portability and taking advantage of the terminal's specific features.

II.7.1 The low-level user interface

The low-level API gives direct access to the terminal screen and events associated with the keys and pointing system. No user interface components are available:

This API includes the Canvas, Graphics and Font classes.

a. The Canvas class

It lets you write applications that can access low-level input events such as :

- terminal keys associated with a code. There are three types of event management method for these keys: "keyPressed()", "keyReleased()" and "keyRepeated()". The latter is not available on all terminals. Use the "hasRepeatEvent()" function to find out whether it is supported.

- the terminal pointer (if one exists). It can be managed by the "pointerDragged()", "pointerPressed()" and "pointerReleased()" methods. To ensure that a pointer is available on the terminal, the "hasPointerEvents()" method has been implemented.

This class offers great control over the display. Games are the best illustration of the type of application that will use this mechanism.

b. The Graphics class

It can be used to produce 2D graphics. It is similar to J2SE's java.awt.Graphics class.

c. The Font class

The Font class represents fonts and their associated metrics.

II.7.2 The high-level user interface

The high-level API provides simple user interface components.

But no direct access to the screen or to input events is allowed. It's up to the MIDP implementation to decide how to represent the components and the mechanism for managing user input.

This API is far richer in classes than the low-level API. In fact, it offers all the classes needed to develop a "classic" interface on this type of terminal. The "List", "ChoiceGroup", "TextBox", "Form", "TextField" and "DatField" classes enable such development.

Other, less conventional classes are also available, such as..:

a. The Alert class

It sets up an alert. This is a dialog box displaying a text message, possibly accompanied by an image or sound. During this display, the user interface is disabled. If a timeout value has been specified, the alert then disappears automatically, otherwise the application waits for a user action.

b. The Gauge class

It defines a gauge that displays a graph in the form of a bar whose length corresponds to a value between zero and a maximum.

c. The Ticker class

A Ticker is a user interface component displaying a line of text scrolling at a certain speed.

d. The Command class

It is used to define a command, the equivalent of the Windows command button.

This class integrates semantic information about an action. It has three properties: the label, the type of command (e.g. return, cancel, validate, exit, help, ...) and the priority level (which defines its location and level in the menu tree).

e. Event management

A high-level event consists of the event source and the event listener. The event comes from the source to the listener, which then processes the event. To implement a listener, the class simply needs to register it with the component from which you wish to listen to events.

There are two types of event:

- the Screen event with its corresponding CommandListener.

- the ItemStateChanged event with the ItemStateListener.

➤ CommandListener

The processing associated with an action performed on a command is carried out in a CommandListener interface. This interface defines a method, commandAction, which is called if a command is triggered.

The corresponding listener is set up by implementing the CommandListener interface. This must then be registered using the setCommandListener (CommandListener myListener) method.

➤ ItemStateListener

Any interactive modification of the state of a form element triggers an itemStateChanged event (e.g.: modification of a text, selection of an item from a list, etc.).

The corresponding listener is set up by implementing the ItemStateListener interface.

We then need to register the ItemStateListener object with a Form form using the setItemStateListener method (ItemStateListener myListener).

II.7.3 Use constraints

It is possible to use both the high-level API and the low-level API in the same MIDlet, but not simultaneously. For example, games that use the low-level API to control the screen can also use the high-level API to display high scores. The low-level API can also be used to plot graphs.

II.8 The lightweight database

RMS (Record Management System) is an API for terminal storage. It's a kind of terminal-independent database, where each record is represented as an array of bytes. Updating involves redoing the entire record.

Records are stored in a so-called Record store. If we want to draw a parallel with relational DBMS, RMS corresponds to the DBMS itself and the Record store to the table. In fact, the parallel to the notion of primary key in relational databases is the "recordID". This is the record identifier. It's an integer. The ID value of the first record is 1, and each new record has an ID value increased by one.

II.8.1 Records store management

There are several ways to manage Records stores.

"openRecordStore" and "closeRecordStore" respectively open and close a Record store. A list of all Record stores can be obtained with "listRecordStore", and "deleteRecordStore" deletes one.

The number of records in a Record store is returned by "getNumRecords". Basic operations on records are performed by these methods: "addRecord", "deleteRecord", "getRecord", "setRecord", "getRecordSize". However, the RMS API has a few additional features for record selection. The first is the use of the "RecordEnumeration" method to list all records in the Record store. The second is the ability to define a filter using the "RecordFilter" method. Finally, the "RecordComparator" interface must be implemented so that records can be compared and sorted.

II.9 The MIDP development cycle
The development cycle for a MIDP application involves the following stages:

- Writing the application with MIDP APIs ;
- Compiling and pre-checking the application ;
- Application testing ;
- Application packaging ;
- Testing the packaged application ;

Given the limited number of classes available in MIDP APIs, MIDlet development is simpler than J2SE application development. We refer here to command-line development.

II.9.1 Writing the application

As with any computer program design, the first step in development is to write the application's source code. In our case, however, certain criteria must be met:

- Each MIDlet must extend the MIDlet class, as with applets, which enables a MIDlet to be started, suspended and terminated.

- A MIDlet must not have a public static void main() method.

It's important to note that when the application is executed, the terminal chooses the location of each command, depending on the type of command and the priority of any other commands **[13]**.

II.10 Conclusion

Java 2 Micro Edition (J2ME) is a technical architecture designed to provide a development platform for embedded applications. The aim is to offer the full power of a language such as Java, combined with the services offered by a restricted version of J2SE.

The J2ME architecture consists of a configuration containing the virtual machine and the minimal functionalities required to access the hardware and network, a profile containing the rest of the functionalities usable on a family of devices with similar characteristics, and optional libraries that manufacturers are free to implement.

We chose J2ME technology.

The next chapter is dedicated to programming Java Bluetooth technology (JABWT).

CHAPTER III
JAVA2APIS FOR BLUETOOTH WIRELESS TECHNOLOGY (JABWT)

III.1 Introduction

Bluetooth is a wireless personal area network technology that enables devices to <u>be</u> linked together without a wired connection, replacing the serial port. The Bluetooth standard is based on a master/slave operating mode. The network formed by one device and all others within its range is known as a "piconet". A master can be connected simultaneously to a maximum of 7 active slave devices. Slave devices have a 3-bit logical address.

This chapter gives a complete description of how to program with JABWT (Java APIs for Bluetooth Wireless Technology) and J2ME. Code samples are provided throughout the chapter, aimed at showing how Bluetooth APIs initialize a Bluetooth application, make connections, install a service, discover neighboring devices and services, then connect to a service (figure III.1).

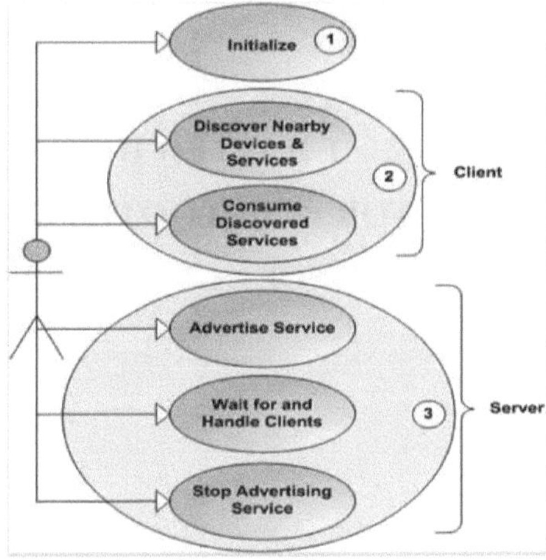

<u>**Figure III.1**</u>: Using a Bluetooth application **[19]**.

- Initialization - This is a mandatory step for a Bluetooth client or server application, as it initializes the protocol stack.

- Customer - *Customer part:* A customer consumes the services that are close to him, if he can find the devices in his vicinity that offer the services he's looking for.

- **Server** _ *Server part*: A server makes services available to clients, registers them and listens to them. Once clients request them, the server accepts and fulfills their requests.

Figure III.2 shows the activities performed by the client and server to establish a connection **[19].**

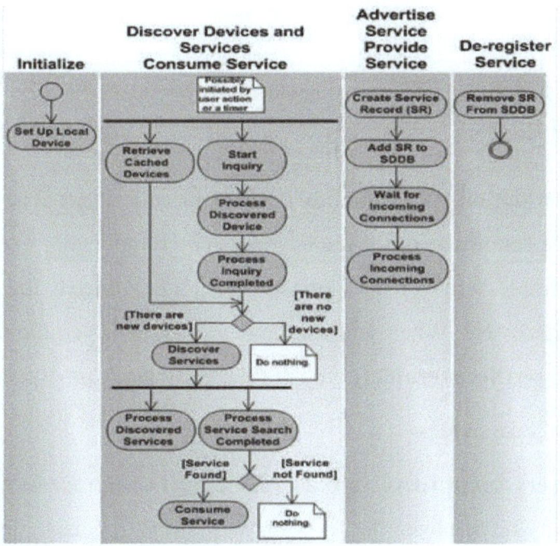

Figure III.2: Client and server activities **[20].**

As shown in Figure III.2, the client and server initialize the protocol stack. The server application prepares services and waits for connections. The client discovers devices and services, then connects to the target device to consume the desired service.

III.2 Initializing the Bluetooth application

Initializing the Bluetooth application is very simple (figure III.3).

First the application looks for a reference to the Bluetooth manager from **LocalDevice.** Client applications look for a reference to the **DiscoveryAgent,** which provides all discoveries related to the server service. Server applications make the device discoverable.

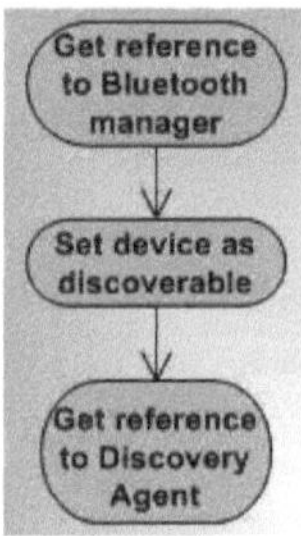

Figure III.3: Initializing the Bluetooth application [20].

You can use an application as a server or as a client. The roles they play depend on the conditions of the application. Server applications set themselves up to be discoverable, while client applications get a reference to the discovery agent for service discovery. When you set the device to discoverable mode by calling **LocalDevice.setDiscoverable()**, you must specify *the inquiry access code (IAC)*.

JABWT's 2 access modes:

> **DiscoveryAgent.LIAC** indicates the limited search access code. The device will only be discoverable for a limited period, precisely one minute. After the limited period, the device automatically returns to undiscoverable mode.

> **DiscoveryAgent.GIAC** indicates the general access code required. There is no limit on how long the device remains in discoverable mode.

To force a device back into non-discoverable mode, simply call **LocalDevice.setDiscoverable()** with **DiscoveryAgent.NOT_DISCOVERABLE** as argument.

III.3 Connection processing

As with all GCF (Generic Connection Framework) connection types, a Bluetooth connection is created using **javax.microedition.io.Connector** from the GCF connection. The URL connection determines the type of connection to be created:

- URL format for L2CAP connection:

  btl2cap ://hostname : [PSM | UUID]; *parameters*

- URL format for Stream RFCOMM connection:

  btspp : //hostname : [CN | UUID]; parameters

Where:

- btl2cap is the URL arrangement for the L2CAP connection.
- btspp is the URL arrangement for the RFCOMM StreamConnection.
- hostname or localhost to set up a server connection, or Bluetooth address to create a client connection
- PSM *is* the Protocol/Service multiplexer value, used by a client connecting to a server. It is similar in concept to a TCP/IP port.
- CN is the channel number value, used by a client to connect to a server, similar in concept to a TCP/IP port.
- UUID *is* the universally unique identifier used when installing a service on a server. Each UUID is guaranteed to be unique across all times and intervals.
- parameters include name to describe the service name, and security parameters authenticate, authorize and encrypt.

  For example:

- A URL of the RFCOMM server:

  btspp://localhost:2D26618601FB47C28D9F10B8EC891363;authenticate=false;encrypt=false;

- An RFCOMM client URL:

  btspp://0123456789AF:1;master=false;encrypt=false;authenticate=false

Using **localhost** as hostname indicates that you want a server connection. To create a client connection to a known device and service, use the service URL found in its **ServiceRecord.**

III.4 Installing a Bluetooth server (figure III.4)

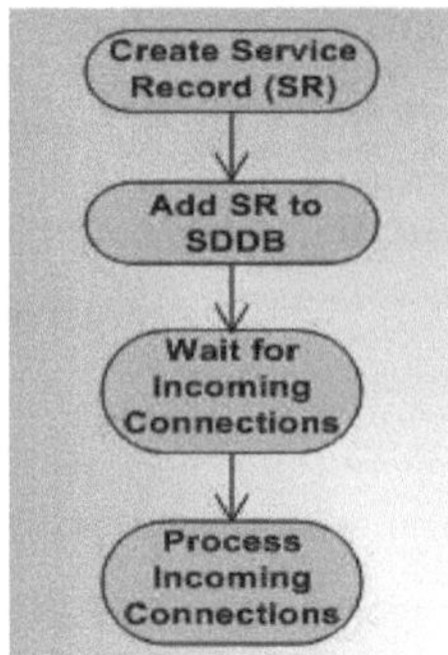

Figure III.4: Bluetooth Server installation **[20]**.

We've installed a Bluetooth server to make a service available for use. There are four main steps:

> By creating a registration service for the service you want to make available.

> Add the new registration service to the Service Discovery database.

> Service registration.

> Waiting for client connection.

Two related operations are significant:

> Modify the registration service, if the service attributes visible to customers need to be changed

> Once this is done, remove the SDDB registration service.

Let's take a closer look at each of these operations.

III.4.1 Creating a registration service

Running Bluetooth automatically creates a registration service when our application creates a notification connection, either a **StreamConnectionNotifier** or an **L2CAPConnectionNotifier.**

Each Bluetooth service attribute has its own unique universal identifier (UUID). You can hardly create a service without first assigning it to the UUID.

III.4.2 Service registration and waiting for incoming connections

Once the notified connection and the **Service** record registration service have been created, the server is ready to register the service and wait for clients. Calling the **acceptAndOpen()** notification method triggers Bluetooth to insert the service record to have a connection associated with the **SDDB**, thus making the service visible to clients. The **acceptAndOpen()** method can block and wait for incoming connections.

When a client connects, **acceptAndOpen ()** returns the connection, in our example a **ConnectionStream**, which represents the client's real goal and allows the server to read the data. The next part of the code waits for and accepts the incoming connection from the client, then reads the Content **[20]**.

This part reads a single **String** from the connection. Different applications **have** different coding requirements. When the **String** can provide dialogue (chat application), the multimedia application can use a combination of character and binary data.

Note that this code blocks when waiting for incoming connections from the client, and must be distributed on its own execution thread.

III.4.3 Updating the registration service

There are cases where the attributes for a registration service need to change. We can update records in the **SDDB** using the direction of a local Bluetooth. As shown in the next section (fragment), we can retrieve the record from the **SDDB** by calling **LocalDevice.get Record ( );** we add or change the attributes of our choice by calling **ServiceRecord; setAttributeValue ( )**, and we transfer the service record back to the **SDDB** by calling LocalDevice .**updateRecord ( ):**

At the heart of Service Discovery is the Service Discovery Database (**SDDB**). The SDDB (figure III.5) is a database maintained by the Bluetooth runtime that contains service records, representing the services that are available to clients **[21]**.

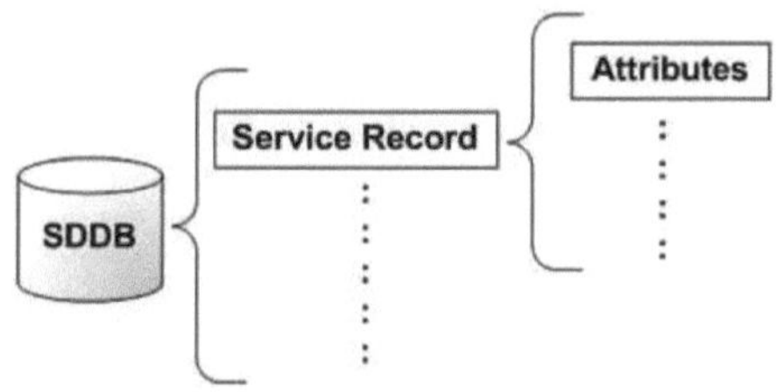

Figure III.5: The SDDB **[20].**

Each service record is represented by a sample ServiceRecord. This record contains the attributes that describe the service in detail. The class provides several useful methods:

- ➢ **getAttributeIDs()** and **getAttributeValue()** search for service disk attributes.
- ➢ **getConnectionURL()** obtains the connection URL for the server hosting the service disk.
- ➢ **getHostDevice()** gets **RemoteDevice()** and services
- ➢ **populateRecord()** and **setAttributeValue()** sets the attributes of the service disk.
- ➢ **setDeviceServiceClasses() sets** the service classes **[20].**

III.4.4 Closing the connection and removing the registration service

When the service is no longer usable, it can be removed from the **SDDB** by closing the notified connection.

streamConnectionNotifier.close();

III.5 Discovering nearby services

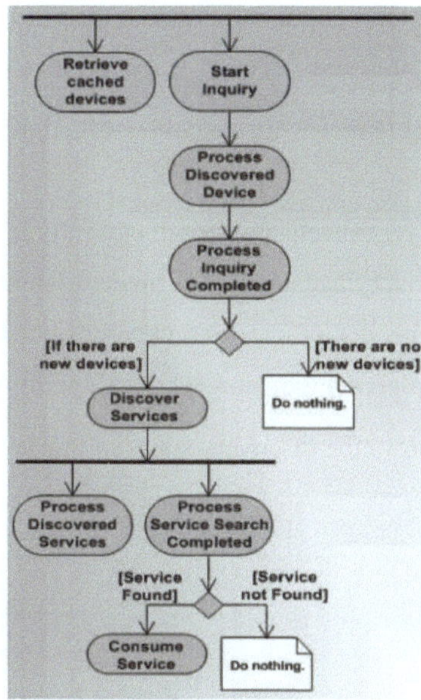

Figure III.6: Discovering devices and services **[16].**

Customers can only use services when they find them. In fact, the search consists of discovering neighboring devices, and then, for each device discovered, searching for the services of choice. In reality, discovering devices is costly and time-consuming.

Discovery is the responsibility of the **DiscoveryAgent**. This class initializes the client and cancels the device discovery service. The **DiscoveryAgent** notifies the client application of any discovered devices and services through the DiscoveryListener interface. Figure III.6 shows the relationship between the client Bluetooth, the DiscoveryAgent and the DiscoveryListener.

III.5.1 APIs for device discovery

The **DiscoveryAgent** method is used to initialize and cancel device discovery.

➢ **retrieveDevices()** searches for neighboring devices already discovered or known.

➢ **startInquiry()** launches the discovery of neighboring devices, also known as search.

➢ **cancelInquiry()** cancels any search currently in progress.

Bluetooth discovery agent calls **DiscoveryListener** device

➢ **deviceDiscovered()** indicates whether a device has been discovered.

➢ **inquiryCompleted()** indicates successful inquiry

Figure III.7 shows the discovery states of the devices.

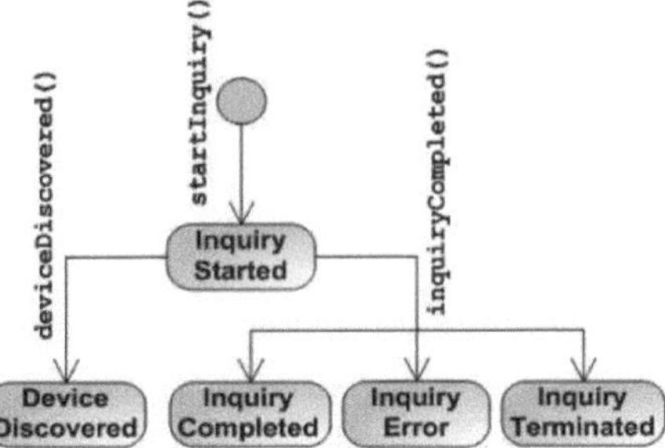

Figure III.7: Device discovery state diagram **[16]**.

Device discovery begins with a call to **startInquiry().** As the search progresses, the bluetooth discovery agent calls **deviceDiscovered()** and **inquiryCompleted().**

III.5.2 APIs for service discovery

We use DiscoveryAgent's service-discovery methods to start and cancel service discovery:

➢ **selectService()** initiates service discovery.

➢ **searchServices()** searches for services.

➢ **cancelServiceSearch()** cancels any service discovery operation currently in progress.

The Bluetooth discovery agent calls DiscoveryListener service-discovery callback methods at various points in the service-discovery phase:

> **servicesDiscovered()** indicates whether any services have been discovered.

> **serviceSearchCompleted()** indicates that service discovery has been completed.

Figure III.8 illustrates the service discovery states reached as a result of DiscoveryListener service callbacks:

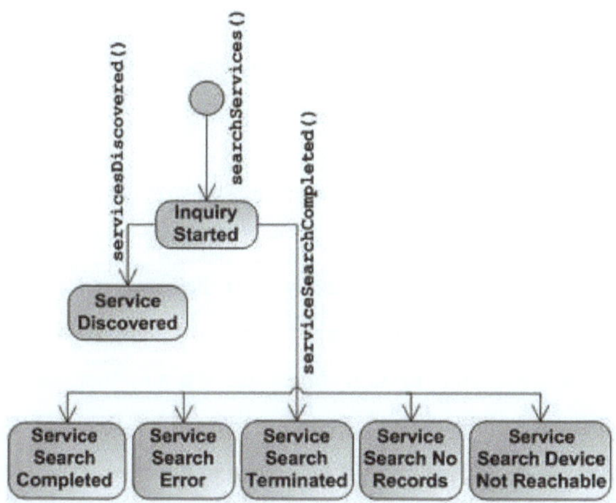

Figure III.8: Service discovery state diagram **[16].**

service discovery begins with a **searchServices()** call, which will result in **servicesDiscovered()** and **serviceSearchCompleted()**

In addition to DiscoveryAgent and DiscoveryListener, we also use the UUID, **ServiceRecord** and **DataElement** classes when discovering services.

III.6 Midlet

In this section we show the structure of a Midlet

```
import javax.microedition.lcdui.Command;
import javax.microedition.lcdui.CommandListener;
import javax.microedition.lcdui.Displayable;
```

```java
import javax.microedition.midlet.MIDlet;
public class MyMidlet extends MIDlet implements CommandListener {
public void startApp() {
}
public void pauseApp() {
}
public void destroyApp(boolean unconditional) {
}
public void commandAction(Command c, Displayable d) {
}
}
```

This **"MyMidlet" Midlet** class provides three abstract methods that the device's application manager uses to communicate with the applications it runs. The **startApp()** method is called immediately after the constructor and every time an application is activated (not just on initial launch). An application can be switched from active to inactive state several times during a single execution, so it's not a good idea to place initialization instructions in it (to be executed only once). This type of instruction should, of course, be placed in the constructor.

The **destroyApp()** method is called by the manager to indicate that an application is about to close. Unlike **startApp(),** this method will only be called once during the execution of an application, so cleanup code can be included.

However, a mobile device is generally less stable than the standard platform and will therefore be regularly switched off or reset by the user. As a result, we can't rely on the **destroyApp()** method being executed.

The third and final abstract method, **pauseApp(),** is used to signal that the application is about to pause. This event occurs when the user switches to another application or uses a device function that prevents the application from continuing to run. As most devices don't have the necessary power to be truly multitasking, care must be taken to free up as many resources as possible within this method. When the application resumes, the handler will call the **startApp()** method **[16].**

The first three methods, **startApp(), pauseApp()** and **destroyApp(),** are required for every **MIDlet.**

During device discovery and service discovery, events will be supplied to. **CommandAction** (), enabled by **CommandListener** interface, is used for event commands **[21]**.

```java
import javax.bluetooth.DiscoveryListener;
import javax.bluetooth.DeviceClass;
import javax.bluetooth.ServiceRecord;
import javax.bluetooth.RemoteDevice;
import javax.microedition.lcdui.Command;
import javax.microedition.lcdui.CommandListener;
import javax.microedition.lcdui.Displayable;
import javax.microedition.midlet.MIDlet;
public class YourMidlet extends MIDlet implements CommandListener,
DiscoveryListener {
public void startApp() {
}
public void pauseApp() {
}
public void destroyApp(boolean unconditional) {
}
public void commandAction(Command c, Displayable d) {
}
public void deviceDiscovered(RemoteDevice remoteDevice,
DeviceClass deviceClass) {
}
public void inquiryCompleted(int param) {
}
public void serviceSearchCompleted(int transID, int respCode) {
}
public void servicesDiscovered(int transID, ServiceRecord[] serviceRecord) {
```

III.7 Conclusion

In this chapter we presented the programming of JABWT java bluetooth technology, which enables devices to link automatically, share data and provide services between them. Sample programs were presented to explain how to use APIs in the development of client-server applications for basic operations, such as client and server initialization, device and service discovery.

This type of programming has enabled us to create our application associated with the laptop screen to display the ECG signal, which we describe in the next chapter.

CHAPTER IV
PRACTICAL IMPLEMENTATION

GOAL

The aim of the project is to transmit an ECG (electrocardiogram curves) to a laptop a few tens of meters away via Bluetooth. The minimum objective is to send at least three leads simultaneously, a lead being a signal that is transmitted by a mobile phone.

analogical view of cardiac muscle activity.

The techniques used will be as follows:

- millivolt voltage acquisition (3 or more channels)

- switched-capacitor filtering

- programming on PIC16F877

- wireless data transmission via Bluetooth.

IV.1 Principle of ECG transmission via cell phone

An important development is the remote transmission of the patient's ECG to the doctor. The doctor is in his office and can diagnose the patient, who remains at home. Transmission is via Bluetooth, then via GPRS (Generate paquete radio service). A cell phone is used as a gateway **[22]**.

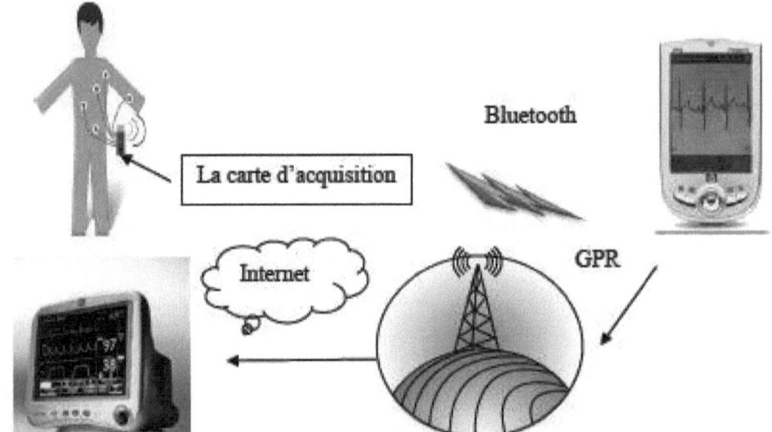

Figure IV.1: Principle of ECG transmission via cell phone

IV.2 Block diagram

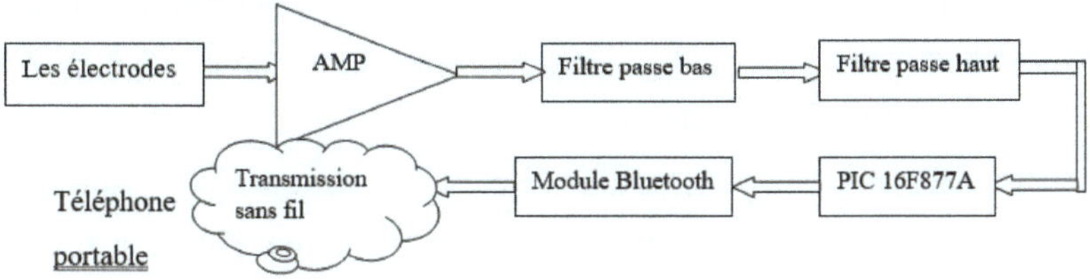

IV.3 Electronic diagram

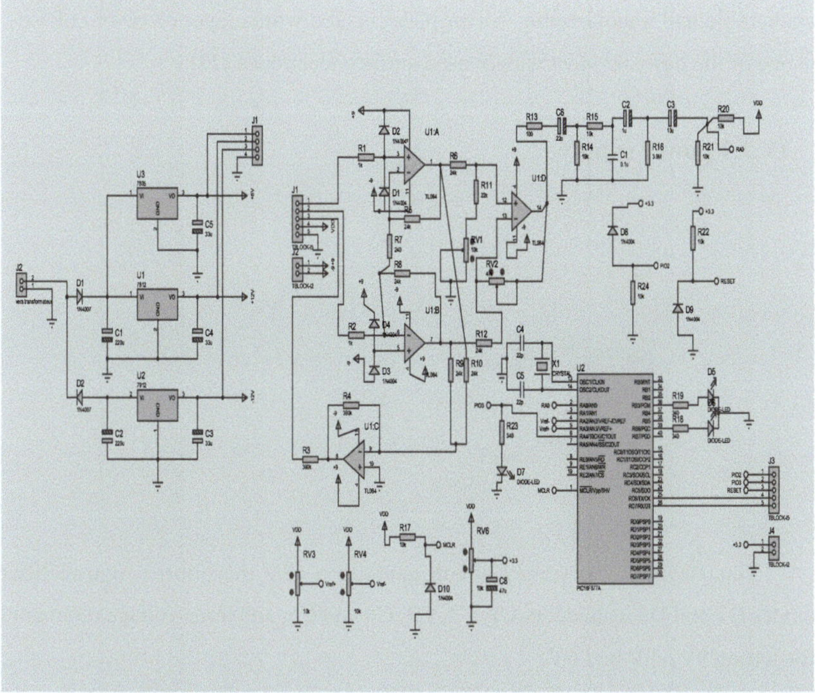

This board has three main stages: the power supply, the ECG detection board and the programming stage for analog-to-digital signal conversion and transmission via Bluetooth.

IV.3.1 Electrodes

The electrode is the first element in the electrophysiological measurement chain, directly in contact with the biological environment. This device detects the heart's electrical activity - the ECG.

The aim is to externally capture the electrical waves emitted by the heart. The sensitive element is therefore a plate made of a conductive material that is placed in contact with the skin.

The important thing about these electrodes is that they are relatively unalterable and impolarizable. Silver plate, coated with a layer of silver chloride, is one of the best and most widely used surface electrodes **[23]**.

IV.3.2 power supply

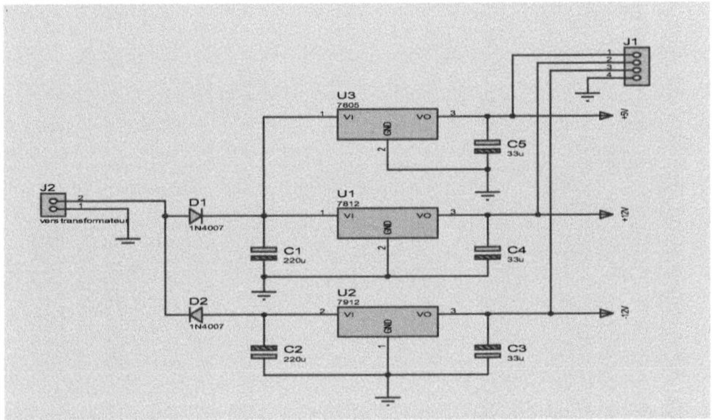

This stable power supply circuit contains a 12V transformer, two rectifier diodes D1 and D2, capacitors C1, C2, C3, C4 and C5, and three voltage regulators for setting 9V, -9V and 5V.

IV.3.3 ECG signal acquisition card

Biological signals are recorded as potentials, voltages and electric field forces produced by nerves and muscles. Measurements involve voltages at very low levels, typically ranging from $1\,\mu V$ to a few mV, high source impedances and superimposed high-level interference signals and noise. It is then necessary to amplify these voltages, to make them compatible with devices such as displays, recorders, or analog-to-digital converters for automated equipment **[24].**

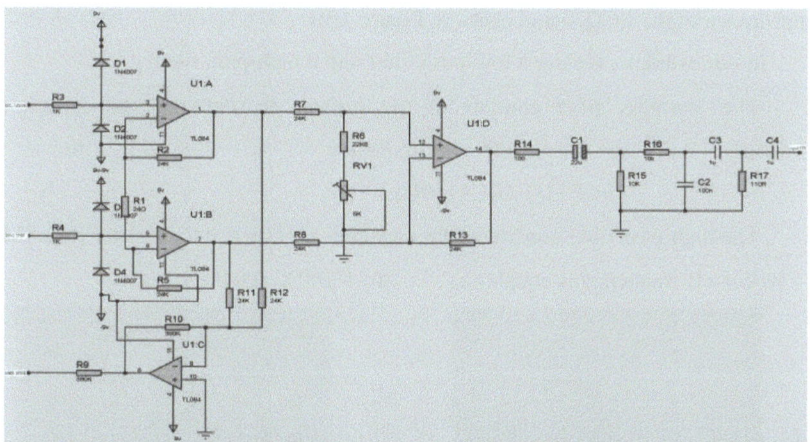

This circuit enables the signal from an electrocardiograph to be displayed on an oscilloscope screen (figure IV.1). The operational amplifiersU1 : A, B and D form an instrumentation amplifier whose gain is 201 according to the following relationship: Avd1= (2R2+R1)/R1. As for U1 : C, it amplifies the common-mode signal by 31x and sends it to the right leg. Thanks to this, the patient's body is brought to a well-defined common-mode level, so that the signal cannot leave the permissible range for the instrumentation amplifier. As a second function, it exerts a feedback effect on the common-mode signal, so that this unwanted signal is further reduced. To protect the inputs from high electrostatic charges, diodes D1 to D4 and resistors R1 and R3 are also used.

The common mode rejection (CMRR) of the instrumentation amplifier can be set by means of P1. To achieve this, start by connecting the two inputs of the same amplifier together. It's essential that the electrodes and skin are in close contact. To test our prototype, we used three bare copper wires, wound on the index fingers (and right leg), which seemed sufficient to provide a good signal. During testing, the amplitude of the ECG signal was 200mV.

Filtering

The ECG signal, amplified in this way, can be drowned in various sources of noise, as shown in Figure 1(a), so a filter is needed to obtain a clearly visualized appearance of the ECG signal peaks in Figure 1(b).

In our example, we use a low-pass filter and a high-pass filter.

The low-pass filter consists of the resistor P=10κΩ and the capacitor C=0.1μF □□sa cut-off frequency is equal to : □□□□□□□□□□□□F=1/2□RC=159Hz

The high-pass filter consists of the capacitor C=1μF and the resistor R=3.9M □, its cut-off frequency is equal to : Φ=1/2πPX=0.04Hz

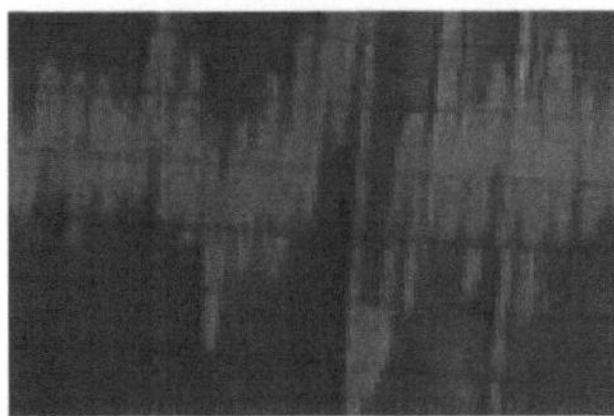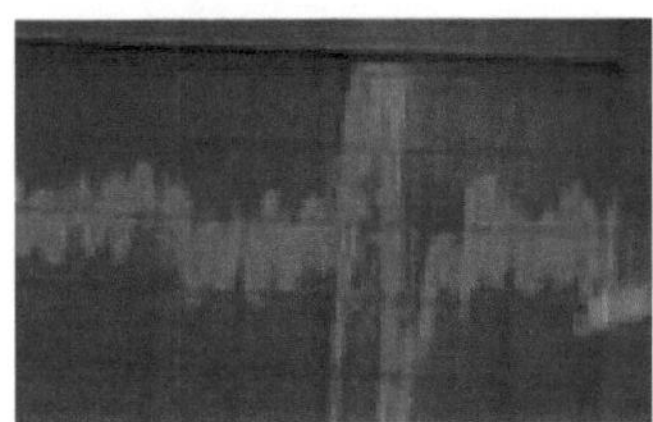

(a) ECG signal before filtering (b) ECG signal after filtering
Figure IV.2: Practical ECG signal visualization.

IV.3.4 CAN board and Bluetooth transmission

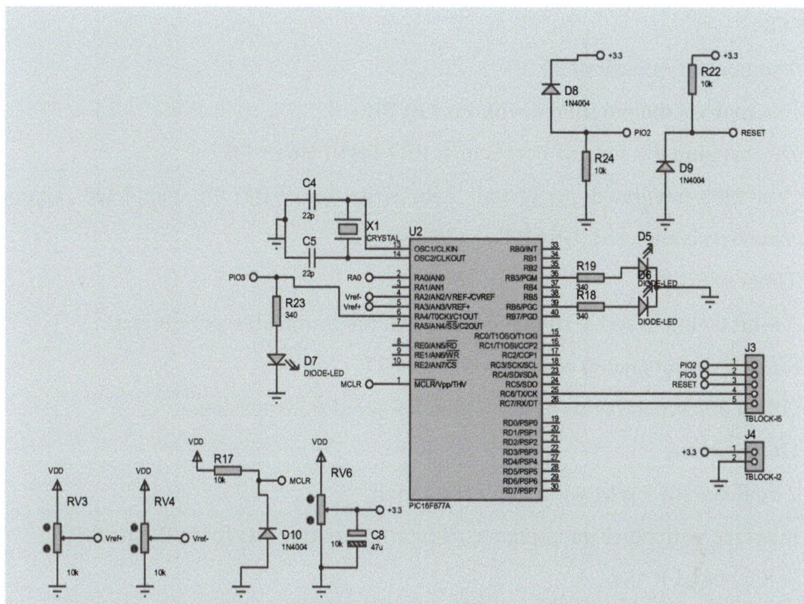

This circuit includes an 8Mhz quartz and C4, C5 capacitors to generate the PIC frequency, two potentiometers one to set PIC Vref(RV3) and the other to give the Bluetooth module supply voltage V=3.3v(RV6), two green and red LEDs to test the supply voltage, another yellow LED to indicate that there is a Bluetooth connection, a voltage divider to eliminate negative values(R20,R21), three push-buttons, two to reset the PIC and Bluetooth module, and the third to disable the Bluetooth module.

a) Programming on PIC16F877A
Description

- Power consumption: less than 2mA at 5V at 4 MHz.

- RISC architecture: 35 instructions lasting 1 or 2 cycles.

- Cycle time: Quartz oscillator period divided by 4, i.e. 500 ns for an 8 MHz quartz.

- Two separate buses for program code and data.

- Instruction code: 14-bit word and 13-bit program counter (PC), allowing 8 K words to be addressed (from h'0000' to h'1FFF')

- 8-bit DATA bus.

- 33 bidirectional input/output ports capable of producing 25 mA per output.

PORTA = 6 bits and PORTB PORTC and PORTD = 8bits PORTE = 3 bits for the

16F877.

- 4 sources of interruption :

- External via the pin shared with Port B: PB0

- By changing the state of Port B bits: PB4 PB5 PB6 or PB7

- Via chip-integrated peripheral: data write to EEPROM completed, analog conversion completed, USART or I2C received.

- Timer overflow.

- 2 8-bit counters and 1 16-bit counter with programmable pre-divider.

- 8-input, 10-bit analog converter for 16F877.

- UART for synchronous or asynchronous serial transmission.

- I2C interface.

- 2 modules for PWM with 10-bit resolution.

- Interface with other microphones: 8 bits + 3 control bits for R/W and CS.

- 368 bytes of RAM.

- 256 bytes of EEPROM Data.

- 8K 14-bit words in Flash EEPROM for the program (h'000' to h'1FFF').

- 1 work register: W and one file register: F for access to the

RAM or to the PIC's internal registers. Both are 8-bit registers.

PORTA: 6 inputs/outputs. 5 CAN inputs. Timer 0 CLK input.

PORTB : 8 inputs/outputs. 1 Clk and Data ext. interrupt input for prog.

PORTC: 8 inputs/outputs. Clk Timer1 and PWM1. USART, I2C.

PORTD: 8 inputs/outputs. Microprocessor interface port (8-bit data).

GATE: 3 inputs/outputs. 3 bits of interf micro control. 3 CAN inputs (see appendix).

This microcontroller is used for three reasons:

-The first is to test the supply voltage of the battery or rechargeable battery by inserting voltage at an analog input.

The second reason is the analog-to-digital conversion of the ECG signal for transmission via Bluetooth, which is coded on 10 bits. This is a value between h'000' and h'3FF'.

The high and low reference voltages can be selected by programming from: VDD or PA3 pin for VREF+ and VSS or PA2 pin for VREF- .

The 4 registers used by the A/D converter module are :

- ADRESH in h'1E' page 0: MSB of the 10 result bits.

- ADRESL in h'9E' page 1: LSB of the 10 result bits.

- ADCON0 in h'1F' page 0: converter control register n°0.

- ADCON1 in h'9F' page 1: converter control register n°1 (see appendix).

-The third reason is serial communication with Bluetooth module M2F03GX/GXA via Tx and Rx pins for transmission.

Transmission is authorized by setting bit 5 of TXSTA to "1": TXEN = 1. The DATA to be transmitted is set in the TXREG register at h'19' page 0. This register warns that it is empty by setting the TXIF flag to "1" (bit 4 of PIR1). This flag changes to "0" as soon as a byte is loaded into the TXREG register. It is reset to "1" by Hard when the register is cleared by transfer to the serialization register: TSR. This register is not accessible by the user, as it has no address. If you then load a 2nd byte into the TXREG register, the TXIF flag will change to "0" and remain there until the TSR register has completely serialized the previous byte to be transmitted. As soon as the STOP of the previous byte has been transmitted, the TXREG register is transferred to TSR and the TXIF flag changes to "1", indicating that the TXREG transmission register is empty and can therefore receive a new byte to transmit. The TRMT bit (bit 1 of TXSTA) provides information on the status of the TSR register. When the TSR register has not finished serializing, TRMT=0. This flag returns to "1" when the register is empty, i.e. when the stop has been issued. The TXIF flag can also be used to generate an interrupt, provided it is authorized by setting bit 4 of PIE1 to "1": TXIE = 1. In this case, peripheral interrupts must be authorized by setting bit 6 of INTCON to "1": PEIE = 1, and by setting bit 7 to "1": GIE = 1 (see appendix).

Programming

The specifications must be translated into an ordered sequence of actions to be performed by the control process. This sequence of operations will be broken down into elementary actions or instructions - the Algorithm.

In our project, we're interested in the PASCAL language, and more specifically in the MikroPascal compiler from Mikroélektronika (version: 8.0.01). We've opted for the MikroPascal compiler because it's easy to master. It also generates hex files. For example, in advanced language, we obtained 2 pages, whereas in assembler we needed 50 pages.

The MikroPascal compiler features a large library of procedures and functions adapted to MICRCHIP's Pic family of microcontrollers. Their use is easily accessible from the software help.

Program organization chart

Before writing the program, we took the specifications for our system, carried out a functional analysis as described above, and then drew up the flowchart shown in the figure, which we translated into MikroPascal language.

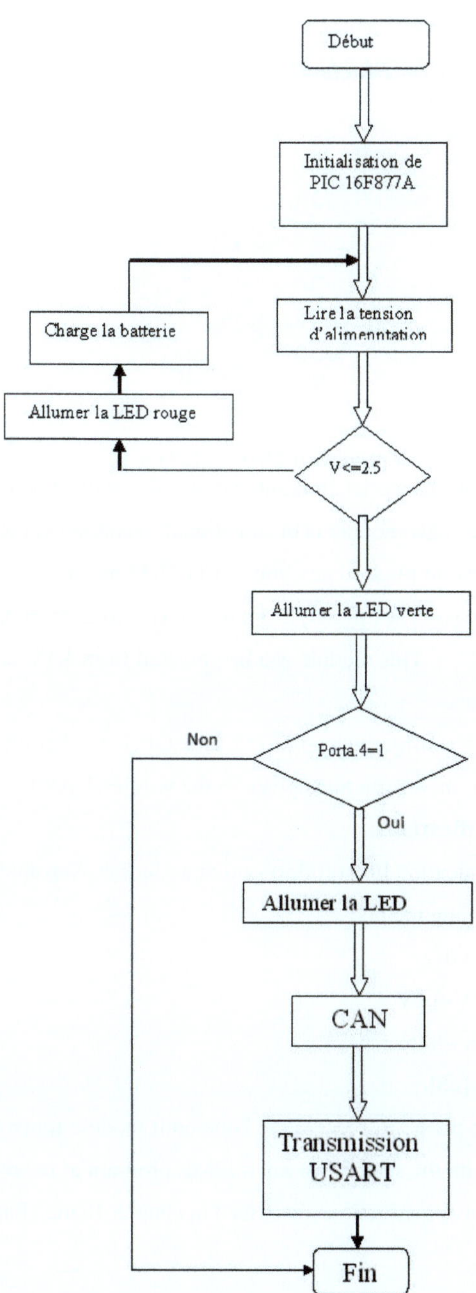

Début
Initialisation de PIC 16F877A
Charge la batterie
Lire la tension d'alimenntation
Allumer la LED rouge
V<=2.5
Allumer la LED verte
Non
Porta.4=1
Oui
Allumer la LED
CAN
Transmission USART
Fin

b) Wireless transmission via Bluetooth
F2M03GX/GXA module

Bluetooth module F2M03GX/GXA

This **module** can replace a serial communication cable with Bluetooth wireless communication. This module is bi-directional, receiving via RX/TX pins and transmitting via transmit pins, so any 9600 to 115200bps serial communication can be passed seamlessly from one device to a target as far away as 100m in the open air!   This module can be powered from 3.1V to 3.6V for simple battery connection.

This module must be configured using F2M_BlueGFG_3_03 configuration software (see appendix), the instructions given in the wirelessUART documentation.   **Specifications:**

- Highly robust communication link (100m) - no more buffer overruns!

- Robust frequency hopping method

- Frequency: 2.4~2.524 GHz

- Operating voltage: 3.1V-3.6V

- Operating temperature: -40 to +80C

- Encrypted connection **[25].**

To enable the cell phone to receive the Bluetooth module transmission and display the ECG signal on the screen, we run a J2ME program to receive the data. This program uses the java instructions discussed in chapter II and chapter III.

Conclusion

In this project, we successfully designed and implemented an innovative system for transmitting electrocardiograms (ECGs) via Bluetooth to a remote cell phone. The main objective was to send at least three ECG leads simultaneously, using sophisticated wireless acquisition, filtering and transmission techniques.

In conclusion, this project represents a significant advance in the field of telemedicine, offering a practical and cost-effective solution for remote patient monitoring. With future improvements and potential scaling-up, our system could have a positive impact on the quality of healthcare and the management of cardiovascular disease.

GENERAL CONCLUSION

In this modest work, the theoretical and practical focus was on diagnostics in the medical field. Other areas were covered, such as wireless transmission to cell phones via Bluetooth. This recent technology is becoming increasingly exploitable in all fields, particularly in the medical sector. Patients can now be monitored remotely by their doctors.

Our circuit was created using Proteus (ISIS & ARES) software from LABCENTER. ISIS enabled us to draw up the electrical diagram and simulate it. ARES enabled us to design the printed circuit.

The numerical part was obtained using the MikroPascal compiler from Mikroelectronika, dedicated to microchip PIC microcontrollers. The latter was chosen for its ease of use. Analog-to-digital conversion was successfully achieved.

The Wireless Toolkit software was also used to simulate cell phone use. Wirelss Toolkit uses the Java (J2ME) language. This part was validated by sending a signal from the server to the client.

This experimental and numerical project enabled us to acquire knowledge of peak programming and Java. We also acquired considerable information on cardiac activity.

It should be noted that, as with any project, we were faced with practical problems. For example, the absence of instrumentation amplifiers meant that we had to use other, more complicated means. These shortcomings have only served to enrich our training and prepare us for the engineering profession.

Finally, we'd like to see our successors tackle arrhythmic signals, trying to trigger alerts, preferably audible, in the event of cardiac anomalies (taquicardiac and bradicardiac). They can also see how information can be sent back via GPRS Mobile-Computer for treatment.

Bibliographies & References

[1] Recommendations concerning the modalities of the medicalized care of the patients in serious condition. SAMU de France, Société française d'anesthésie réanimation,
July 2002.
[2] Garrigue B. Management and vigilance of medical devices in SAMU/SMUR. *In* 11th IADE Advanced Training Meeting. Paris, Elsevier 2000.
[3] Isaac NEWTON "ECG signal modeling with orthogonal polynomials" http://www.ECG.compréssion.polynome.pdf .
[4] http://www.iav.ac.ma/veto/filveto/guides/phys/physiopharmazine/physiologie cardiac.htm#HISTOLOGY.
[5] http://www.chapitre1 heart and electrocardiography.
[6] http://membres.lycos.fr/tpecardio/ecg.htm.
[7]
[8] http://www.introduction-sommaire.pdf.

[10] J. LE ROUX, *"Notion de communication numérique"*, March 20, 2001. http://www.essi.fr/~leroux/courstransmission.pdf.
[11] http://www.dotnetguru.org/articles/J2MEvsSDE/J2MEvsSDE.htm .
[12] Bruno Delb, J2ME Java Applications for Mobile Terminals, Editions Eyrolles, 2003.
[13] Martin de Jode, "Programming Java 2 Micro Edition on Symbian OS A developer's guide to MIDP 2.0" ,John Wiley & Sons, lTD, 2004.
[14] Ian Utting, "Problems in the Initial Teaching of Programming using Java: The case for replacing J2SE with J2ME", ITiCSE'06, Bologna, Italy.Ian Utting, ACM, June 26-28, 2006
[15] Damien Zéni, "Présentation personnelle Java 2 Micro Edition", EIVD - EI5b, *February 2003.*
[16] André N. Klingsheim, J2ME Bluetooth Programming Master's Thesis, Department of Informatics University of Bergen, 30th June 2004.
[17] Jean Michel DOUDOUX, "Développons en Java", Electronic book, http://jmdoudoux.developpez.com/java/.

[19] C. Enrique Ortiz, "Using the Java APIs for Bluetooth, Part 2 - Putting the Core APIs to Work", http://developers.sun.com/techtopics/mobility/apis/articles/bluetoothcore/index.html
[20] Kumar, "Bluetooth Application Programming with the Java APIs", First Edition, Morgan Kaufmann, 2004.
[21] C. Enrique Ortiz, "Using the Java APIs for Bluetooth", Part 1 API Overview, http://developers.sun.com/techtopics/mobility/apis/articles/bluetoothintro/index.html

[22] http://www.rds_2004.pdf.
[23] donfack Colince electrodes; *"characterization of electrode_tissue contacts for implantable neuromuscular stimulators".*
[24] Memoire présentée en vue de l'obtention du diplôme de maitrise des sciences appliqués en genie électrique Ecole polytechnique de montréal. January 2000.
[25] http://www.datasheet F2M03GX/GXA.pdf.

Printed by Books on Demand GmbH, Norderstedt / Germany